Table of
CONTENTS

CHAPTER ONE

What Is a Blog?

Imagine your family has been chosen to help a puppy become a **seeing-eye dog**. You will be taking care of and training the puppy. That puppy will grow up to help someone live a more independent life. All of your family's friends keep asking how Mimi the puppy is doing. How is she growing? What kinds of helping tasks has she learned?

Create and Share | Thinking Digitally

Building a Blog

By Kristin Fontichiaro

Published in the United States of America by Cherry Lake Publishing Group
Ann Arbor, Michigan
www.cherrylakepublishing.com

Series Adviser: Kristin Fontichiaro
Reading Adviser: Marla Conn, MS, Ed., Literacy specialist, Read-Ability, Inc.
Book Designer: Felicia Macheske
Character Illustrator: Rachael McLean

Photo Credits: © BestPhotoStudio/Shutterstock.com, 5; © arrowsmith2/Shutterstock.com, 9; © ilkercelik/Shutterstock.com, 13; © Andrey_Popov/Shutterstock.com, 15; © Sunny studio/Shutterstock.com, 17; © Picsoftheday/Shutterstock.com, 21

Graphics Credits Throughout: © the simple surface/Shutterstock.com; © Diana Rich/Shutterstock.com; © lemony/Shutterstock.com; © CojoMoxon/Shutterstock.com; © IreneArt/Shutterstock.com; © Artefficient/Shutterstock.com; © Marie Nimrichterova/Shutterstock.com; © Svetolk/Shutterstock.com; © EV-DA/Shutterstock.com; © briddy/Shutterstock.com; © Mix3r/Shutterstock.com

Cherry Lake Press is an imprint of Cherry Lake Publishing Group.

Library of Congress Cataloging-in-Publication Data has been filed and is available at catalog.loc.gov

Cherry Lake Publishing Group would like to acknowledge the work of the Partnership for 21st Century Learning, a Network of Battelle for Kids. Please visit *http://www.battelleforkids.org/networks/p21* for more information.

Printed in the United States of America
Corporate Graphics

What kinds of pet stories could you tell?

You sound like you need a **blog**. The name comes from combining the words "web" and "log." It is an online journal or diary that you write. A blog is made up of a series of **posts**. You might write a new post each day or week. Or maybe you just write when Mimi learns something exciting.

You can write blog posts by yourself. Or you can take turns writing them with other students or family members. People who find your blog can read what you have to say. Then they might leave you notes called **comments**.

Blogs can be a fun way to keep track of how things are changing around you. With the help of your family or teacher, you can start one for free.

ACTIVITY

Choose a Blogging Platform

The online service you use to publish your blog is called a platform. Your teacher can help you set up this special site. Teachers might use a tool like Kidblog to set up blogs for you and other students. Your parents can also set up a blog using a platform like WordPress.com or Blogger.com. They can set it up in their name and share it with you. You have to be at least 13 to set up your own public blog. Look over a few platforms. See which has designs and features that you like. Start thinking about what your blog will be about and what you will name it.

CHAPTER TWO

URL and Title

All blogs have a URL. That's the web address people type in to find your blog, like https://mimitheguidedog.blogspot.com or https://mimitheguidedog.wordpress.com. (These are example websites. They don't exist.) When you set up a blog, you will need to create a URL. Have a few in mind in case your first choice has already been claimed by someone else.

Next, you will need to give your blog a header. This is the area that goes across the top of your blog. It can just be the **title** of your blog. Or it can also include a **graphic**, or a decorative design.

Your blog's URL and title should match what you write about.

Sometimes the title is the same as the URL, like "Mimi the Guide Dog." But it can also be different, like "Our Amazing Adventures Teaching Mimi to Be a Good Guide Dog Instead of a Naughty Puppy." You can add a subtitle, which is a few words that tell more information about your blog. A subtitle for your blog could be "Our Family's Journey to Save Our Shoes from Being Eaten by a Young Dog."

Your blog is a personal web page. But be careful with the amount of information you share online. You want to protect the privacy of your family. For example, it's safe to share Mimi's name. But it's a good idea to leave off your last name, your family's address, your school name, and other personal information.

ACTIVITY

Claim Your Blog

Work with your teacher or a trusted adult to set up your blog using one of the platforms in chapter 1. Have some possible URLs and titles ready to go. Remember to keep private information out of your URL, title, and subtitle. Store your username, password, and URL in a safe place.

CHAPTER THREE

Designing Your Blog

This is a really fun part of starting a blog. After you log into your blog, you can choose a **theme**. Themes help organize and set the mood of your blog through designs. Themes help you choose colors and **fonts** and how the **content** is arranged on the page. You will need to **upload** a graphic or a photo for your blog header. Or you can simply have your blog title act as your blog header.

Some blog **layouts** have sidebars. These are long vertical boxes of information on the left or right side of your blog posts. Work with an adult to figure out what information to put there. Some people list **links** to their friends' blogs. Some put photos or quotes there. You can also put other information in the footer, at the bottom of the page. The fun part is that you can change your layout whenever you want.

For inspiration, check out a few blogs started by kids.

When someone types in your URL, they will see your most recent post. How much they see will depend on how you set up your blog. They might see the entire post. Or the home page might just show a few lines or even just the title. In that case, readers would click to view the entire post. Blog posts are always organized with the newest items first.

You have the option to turn on the commenting feature for your blog. This lets readers leave you a note about what you have written. You can decide if you want their note to show up on your site or not.

ACTIVITY

What's Your Blog's Mood?

You will be surprised at how many design choices you have when creating a blog. Now is a good time to think about your blog's personality or mood. What colors will you choose? Do you like fancy fonts that look like invitations? Do you want your readers to see a lot of text all at once? Or will they just see the blog post titles? Your design choices will determine how people feel when they arrive at your site. Think about Mimi's blog. Write down as many words as you can that describe the mood you would design for it.

CHAPTER FOUR

Let's Get Writing

You're ready to write your first blog post! Let's write a post in which you introduce Mimi to the world.

You'll need a post title and some content. This is different from a blog title. A blog title is like the title of a book. The post title is like the chapter titles in that book. It should tell the reader something about what is going to happen in the post. Type what you want to say in the body of the post. You can add photos or even a video.

What would you write about?

SAMPLE POST

Welcome Home, Mimi the Guide Puppy!

Posted on September 4, 2020

Today was a special day at our house. We brought Mimi home from the Leader Dogs for the Blind organization. We are going to take care of her and help her learn how to be a good guide dog. Mimi is a golden retriever with soft, yellow hair. She wears a special vest that says "Guide Dog" and a harness with a handle so that we can hold on to her and feel her by our side. She does not use a leash. A leash would let her be too far away from her owner to be helpful. We will not keep Mimi forever because she will become the special helper of someone who is blind. Mom says we will love her while she is here.

Here is what she looks like today. How big do you think she will be when she leaves us?

When you are done writing your post, click Publish. Then people will be able to see your post. Congratulations! You've written your first blog post.

It's fun to tell a blog author what you think. When you comment, keep these things in mind:

- Tell people what you liked. Say, "Thanks for showing us what your new puppy looks like!" not, "That was a terrible picture."

- Help the writer know what they did well or could do better. Say, "I like how you described her fur. It made her seem real!" or "Thank you for sharing! Can you tell us whose room she sleeps in?"

- Be specific. Saying "Good job" sounds like you are just being kind. What was good? The design? The word choice? The photos? "Great close-up on Mimi's face!" is specific.

ACTIVITY

Post Titles

Imagine that you are ready to write a few posts about big moments in Mimi's life. What titles might you give these blog posts?

Topics:

- We took Mimi to the groomer to get her nails trimmed. She did not like it at all.
- Mimi got a lesson on how to help her person know when it is safe to cross the road. She did a very good job.
- Mimi's trainer came to visit us today. He taught Mimi how to turn a doorknob. We couldn't believe she could make her paws turn the knob!

GLOSSARY

blog (BLAWG) a website that acts like an online journal

comments (KAH-mentz) notes that readers leave about your blog posts

content (KAHN-tent) the words, images, and videos that you put into your blog post

fonts (FAHNTS) styles of printed letters and numbers

graphic (GRAF-ik) an image, photo, logo, icon, or design

layouts (LAY-outs) where everything is placed on a web page

links (LINGKS) pieces of text on a web page that connect to another web page when clicked

posts (POHSTZ) the articles you write and put on your blog

seeing-eye dog (SEE-ing EYE DAWG) a dog trained to lead people who can't see

theme (THEEM) the particular subject or idea on which the style of something is based

title (TYE-tuhl) the name of your blog or of an individual blog post

upload (UHP-lohd) to send information to another computer over a network

For More INFORMATION

BOOKS

Minden, Cecilia. *Writing a Blog*. Ann Arbor, MI: Cherry Lake Publishing, 2020.

Orr, Tamra. *Blogging*. Ann Arbor, MI: Cherry Lake Publishing, 2019.

WEBSITES

Kidblog
kidblog.org/home
This site lets teachers set up blogs for their students.

YouTube—What Is a Blog?
https://www.youtube.com/watch?v=NjwUHXoi8lM
Get a quick tour of how blogs work and are set up.

INDEX

About the AUTHOR

Kristin Fontichiaro teaches at the University of Michigan School of Information and writes books for adults and kids.